Housing
86

蜜蜂的数学
The Mathematics of Bees

Gunter Pauli

〔比〕冈特·鲍利　著

〔哥伦〕凯瑟琳娜·巴赫　绘

唐继荣　译

上海远东出版社

丛书编委会

目录

Contents

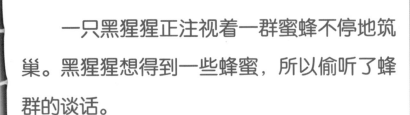

　　一只黑猩猩正注视着一群蜜蜂不停地筑巢。黑猩猩想得到一些蜂蜜，所以偷听了蜂群的谈话。

　　"你们吃了太多的蜂蜜，却没有生产足够的蜂蜡。"蜂王斥责道。

$\mathcal{A}$ chimpanzee is watching a swarm of bees working away at their hive. He wants some of their honey so is eavesdropping on their conversation.

"$\mathcal{Y}$ou are eating too much honey and not making enough wax," scolds the queen bee.

你们吃了太多的蜂蜜

you are eating too much honey

更多蜜蜂扇动他们的翅膀

More bees to flap their wings

"抱歉，陛下，"一只工蜂回答道，"我们正竭尽全力挤压出蜡片，用来建造蜂巢。"

"为了让温度保持在35摄氏度，我们还需要更多蜜蜂扇动他们的翅膀。"蜂王坚持道。"这是我们能让蜂蜡保持坚固，但仍具可塑性以用于筑巢的唯一方式。"

"I apologise, Your Majesty," responds one of the worker bees. "We are doing the best we can to squeeze out flakes of wax to make the honeycomb."

"We also need more bees to flap their wings to keep the temperature at exactly 35 degrees Celsius," insists the queen. "This is the only way we can keep the wax firm but still workable enough to build the hive."

"那么，安排更多的蜜蜂进到蜂巢里来吧，越多越好！"

黑猩猩偷看到蜂巢里有无数形状完美的六边形小窗口。这些蜜蜂不欢迎他的闯入，因此他在被蜜蜂蛰伤前迅速离开了。

"每间蜂室的室壁必须以正好120度的夹角相交，这样我们才能建造出这种六边形的室壁。"一只工蜂喊道。

"So let's get more bees into the hive, the more the merrier!"

The chimpanzee peeks inside and sees perfectly formed, six-sided clusters of little windows. The bees do not like his intrusion, so he quickly moves away, before he gets stung.

"The walls of every cell must meet at exactly 120 degrees so that we can build this wall of hexagons," shouts a worker bee.

正好120度

Exactly 120 degrees

永远不要把室壁的厚度造得超过头发丝的直径

Never make the wall thicker than a hair's width

"我们还必须节约材料！"另一只工蜂说。"永远不要把室壁的厚度造得超过头发丝的直径。"

"如果我们正确地建造好蜂巢，那么我们采集的花蜜将永远不会滴出，我们就有足够的蜂蜜来养活我们的孩子了。"工蜂说。

"And we must save on materials!" says another. "Never make the wall thicker than a hair's width."

"If we build the hive properly, the nectar we collected will never drip out and we can have enough to feed all our babies," says the worker.

"这就是我们要把蜂巢的蜂室倾斜的原因，这样储存的蜂蜜才不会滴出来。"

"当我们只用六边形的窗口时，我们能在蜂巢中储存更多的蜂蜜。如果我们用三角形或四边形的窗口，那就不能像这样储存蜂蜜了。"

"That's why we tilt the cells of our honeycomb so that the stored honey will not drip out."
"We can keep more honey in our hive when we only use six-sided windows. It would never work if we were to have triangular or square-shaped windows."

把蜂室倾斜

Tilt the cells

正方形或长方形的砖瓦

Squares or rectangular tiles

"为什么人类总是用正方形或长方形的砖瓦、墙壁和窗户呢？"一只蜜蜂好奇地问。

"因为他们从未真正地认真对待数学。"

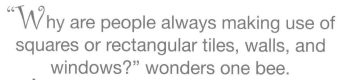

"Why are people always making use of squares or rectangular tiles, walls, and windows?" wonders one bee.
"Because they never really took maths seriously."

"但人类擅长研究数学，他们甚至开发出能为你运算一切的计算机程序。"

"擅长？"一只工蜂惊讶地问。"他们知道的数学是线性的，而自然界中的绝大部分事物都不遵循这样的规律。我们的六边形窗口是一个例外。"

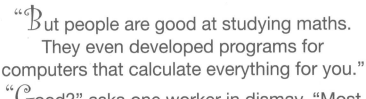

"But people are good at studying maths. They even developed programs for computers that calculate everything for you."
"Good?" asks one worker in dismay. "Most maths they know is linear and hardly anything in nature follows that logic. We, with our six-sided windows, are one of the exceptions."

能为你运算一切的程序

Programs that calculate everything for you

气泡、泡沫和水晶怎样形成

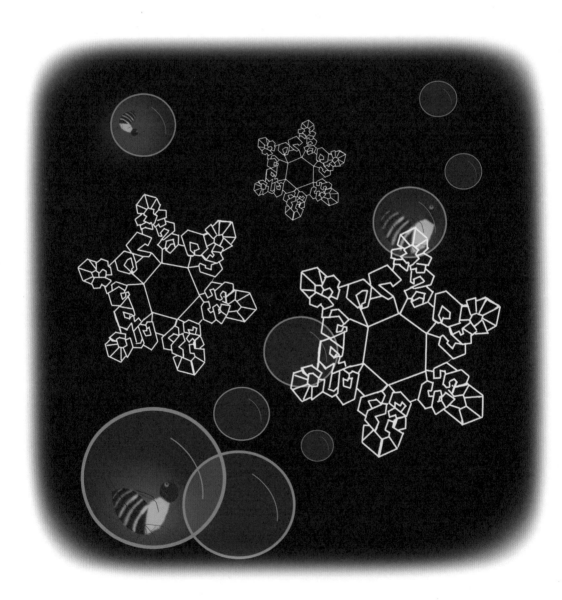

How bubbles, foam, and crystals form

"那么，当人类连气泡、泡沫和水晶怎样形成，以及植物细胞为什么从不沿直线生长都不懂时，为什么他们还想去研究基因？"

"这是因为太多人不知情。或许他们中的一些人喜欢这种状态，满足于他们所知道的一切。"

"So why do people want to play around with genes, when they don't even know how bubbles, foam, and crystals form, and why plant cells never grow in a straight line?"

"It is because so many people are uninformed. Perhaps some of them prefer it that way and want to stay in the comfort zone of what they know."

"因此，这些智人更应该被称为'不智之人'！"

　　"好吧，让我们相信他们的孩子将会做得更好一些吧。"蜂王叹息道。

　　……这仅仅是开始！……

"So these Homo sapiens should rather be called Homo non sapiens!"

"Well, let's trust that their children will do better," sighs the queen.

… AND IT HAS ONLY JUST BEGUN!…

... AND IT HAS ONLY JUST BEGUN! ...

Did You Know?

你知道吗?

Honeybees produce their own antitoxins to break down more than 100 kinds of pesticides and chemical treatments against mites. When commercial bees are fed high fructose corn syrup (HFCS), they are no longer able to do this.

蜜蜂生产自己的抗毒素，能分解超过 100 种农药，还能生产抗螨虫的化学物质。当商用蜜蜂被喂食果葡糖浆（HFCS）时，它们将不再具备这种能力。

Bees visit 600 flower sites and pollinate over 40 crops every season. Commercial bees' ability to pollinate is negatively affected when fed HFCS containing minute amounts of chemicals (20 parts per billion), which is well below permitted concentrations.

蜜蜂每个季度访问 600 处开花点，为 40 多种农作物传粉。当喂食的果葡糖浆含有微量人造化学物质（20ppb）时，即便远低于允许浓度，蜜蜂的授粉能力仍然会受到负面影响。

烟碱类农药对蜜蜂来说是最毒的农药。农药应用于一般植物、种子、几乎所有形式的转基因玉米，还有花椰菜、西蓝花、苹果和梨。

$\mathcal{N}$eonicotinoids are the most toxic pesticides for bees. Pesticides are applied to plants, seeds, and on virtually all forms of genetically modified corn, but also to cauliflower, broccoli, apples, and pears.

与那些生活在蜂巢中的蜜蜂相比，为获取食物而经常探索新环境的蜜蜂有一种不同的大脑遗传活动。它们大脑中的化学信号与寻求刺激的人的大脑情况相似。

$\mathcal{B}$ees that constantly explore new environments for food have a different form of genetic brain activity than those that live in hives. The chemical signals from the brain are similar to those in thrillseeking people.

无刺蜂并不温顺，它们仍然会攻击：雄蜂在半空中撞击，雌蜂叮咬敌人，直至一方跌落地面死亡。这种战术用于保护自己，以及入侵其他蜂巢以传播基因。

蜜蜂能选择收集高蛋白的花粉，还是富含糖分的花蜜。花蜜是糖类丰富的来源，也是制造蜂蜜的原料，而花粉是蜜蜂幼虫的高蛋白食物。

蛋白质

蜂蜜

Chimpanzees living in dense forests in Central Africa use as many as five different self-made tools to get honey from beehives located up to 20 m up a tree and up to 1 m underground.

黑猩猩生活在非洲中部地区的茂密森林中，能使用5种不同的自制工具来采食蜂巢中的蜂蜜，而这些蜂巢上至20米高的树上，下至地下1米。

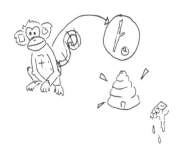

Bees have been taking collective decisions for 30 million years. In a swarm of 10 000 bees there are 300-500 bees that join the search for new nest sites over areas large as 30 square kilometres.

3 000万年来，蜜蜂一直采用集体决策的方式。在多达1万只蜜蜂的蜂群里，有300—500只蜜蜂加入寻找新巢址的队伍，探索范围达30平方千米。

30 平方千米

If bees pollinate apple and broccoli flowers and the insecticides used to protect these crops kill bees, should we continue using insecticides?

如果蜜蜂为苹果和西蓝花的花传粉，而用于保护这些农作物的杀虫剂会杀死蜜蜂，我们还应该继续使用杀虫剂吗？

蜜蜂是聪明的建筑师，高效利用空间和建筑材料。我们人类建筑师是否应该学习蜂巢背后的工程学和数学呢？

Bees are smart architects, using space and building materials efficiently. Should architects study the engineering and mathematics of beehives?

Bees are not only architects; they are also very efficient in working together in large numbers. Could managers of companies learn anything interesting from them?

蜜蜂不仅是建筑师，而且在群体共同工作时也非常有效率。公司里的经理们能否从它们那里学到一些有趣的东西？

除了六边形之外，蜜蜂从来不在它们的建筑中采用直线，那么为什么人类常用直线和 90 度夹角呢？

If bees never use straight lines in their buildings, except for the hexagonal tile, why do people mostly make use of straight lines and 90° angles?

Triangles, squares, and hexagons are some of the only geometric shapes that can be used in tiling. Cut out some cardboard tiles, at least 20 of each shape, and try to cover a sheet of paper with them. Calculate how many tiles you will need of each of these geometrical shapes to cover a sheet of paper. Which shapes were the easiest to make and which ones were easiest to put together? After doing this on a flat, two-dimensional surface, see if you can glue the tiles together to make a three-dimensional beehive shape.

　　三角形、正方形和六边形是仅有的能用于砖瓦的几何形状。剪出一些纸板，每种形状至少 20 块，设法用它们来盖住一张纸。对于这些形状的纸板，计算每种形状需要多少块才能盖住一张纸。哪种形状最容易剪，哪种形状最容易拼在一起？在平坦的二维表面完成这项工作后，你能否把这些纸板用胶水粘在一起，用来制作一个三维的蜂巢状物体。

学科知识

Academic Knowledge

生物学	蜜蜂依赖储存的花粉越冬，使冬季成为净化身体的"节食期"；蜜蜂通过摇摆运动来交流；雄蜂只进行交配，并不采集花蜜或传粉，当食物缺乏时常被从蜂巢中赶出来；蜂王寿命长达5年；蜜蜂身体完全被毛覆盖，甚至眼睛上也是如此，以此最大可能地采集花粉。
化 学	烟碱类物质被广泛用作杀虫剂，这也是蜜蜂减少的原因之一；对蜜蜂来说，农药降解产物通常比农药原来的化学物质毒性更大；蜂蜜含有天然的防腐剂，因而长久不变质；碳环结构中六边形是最稳定的结构，而石墨由类似蜂巢的六边形碳原子网络结构组成。
物 理	圆形的面积周长比比六边形的更大，但当你用圆形砖瓦时，空间会浪费；蜜蜂扇动翅膀来为蜂巢加热，使已消化的花蜜中的液体蒸发来生成蜂蜜；蜜蜂也可扇动翅膀来为蜂巢降温。
工程学	蜜蜂建造蜂巢时，从制作粗糙的圆形小室开始，因为这对材料的利用最充分，然后将这些材料弯曲成六边形；幼蜂分泌圆珠笔尖大小的蜡片，其他的蜂则生产厚度不超过0.2毫米的圆柱形腔室。
经济学	商用蜜蜂在授粉方面的效率不如野蜂，这证明野生昆虫不能被替代；据估计，昆虫授粉对世界经济的贡献每年超过2 000亿美元。
伦理学	一些人缺乏学习能力，而另一些人却拒绝学习；某些科学家希望能批准进行遗传操作，将人类遗传物质与合成化学物质结合，但他们并未意识到这样做将对生态系统造成冲击。
历 史	埃及人从4 500年前便养蜂；公元4世纪的几何学者帕普斯提出，蜜蜂选择六边形为蜂室的形状，是因为这样只需使用最少量的蜂蜡。
地 理	除了在植物只能自花授粉或依赖鸟类传粉的南极洲外，其他所有大洲都有蜜蜂分布。
数 学	蜜蜂需要3—4千克蜂蜜才能生产出1千克的蜂蜡；为了生产一茶匙的蜂蜜，一只蜜蜂必须往返花丛150次，而为了生产1 000升的蜂蜜，需要造访花丛400万次。
生活方式	虽然一些人接受转基因食品，但许多人并不接受这种食物，而是更愿意接受他们本地的文化、传统和生态系统。
社会学	蜜蜂采取集体决策方式已长达3 000万年，这有助于它们的生存；蜜蜂被称为社会性昆虫，因为它们生活在群体中，相互依赖；"嘴唇抹了蜂蜜"是描述口才好的一种表达方式。
心理学	持之以恒：耗费多年精力证明六边形是最有效的储存模式。
系统论	为了可持续性，我们必须用一个系统来替代另一个系统，而不是用一件产品来替代另一件产品，或者用一个过程替代另一个过程。

情感智慧
Emotional Intelligence

蜜　蜂

蜜蜂很热情，行为举止彬彬有礼，并且勇于认错，竭尽所能把事情做得最好。他们作了很多努力，并忠诚于集体。他们的工作精度很高，体现了他们的非凡技能，以及掌控环境的能力。他们欢迎其他成员加入他们的群体，并分享他们的知识和目标：建设蜂巢、找到食物。蜜蜂保卫蜂巢和后代，决心攻击任何入侵者。这证明体形不是最重要的方面，蜜蜂能依赖他们庞大的个体数量来对付所有的敌人。蜜蜂执着于资源效率，不纵容浪费，并懂得对几何学的机智利用将提高他们的效率。蜜蜂将他们的数学与人类的数学进行对比，得到的结论是：人类肯定是无知的。然而，他们希望人类的下一代能做得更好。

艺术
The Arts

找一些蜜蜂的特写图片，研究它们以及蜂巢六边形的窗口形状。想象一下你能用哪些方式将你看到的整合进一件艺术作品。你可以从生物学家安娜·瓦尔马（Anna Varma）那里得到启发，她拍摄了一些令人惊奇的蜜蜂及其生活条件的照片。现在画一幅包含六边形与蜜蜂一家的图画，你得花时间才能让你的绘画反映出蜜蜂的忙碌。

思维拓展
Systems: Making the Connections

蜜蜂是生态系统中的重要角色，是生态系统健康非常重要的指示器。蜜蜂已经被许多文化所赞美，如玛雅、中国、埃及和古希腊文明。蜜蜂在社会组织结构、资源效率和物理建筑结构上具有令人印象深刻的指挥能力，它还有非凡的免疫系统，从来不生病。蜂蜜、蜂蜡和蜂王浆总是被用于治疗。最近蜜蜂开始遭受睫毛上螨类的侵害，但很快就弄清楚这是由蜂巢中人造化学物质积累引起的生理应激所致。即便蜜蜂拥有自己的抗毒素，但随着时间的推移，微量但多种多样的人造化学物质，以及食物从天然食物（花蜜和花粉）到果葡糖浆的转变，导致世界上80%的蜜蜂种群崩溃。野蜂没有能力为上千公顷的植物授粉，而被喂食果葡糖浆的商用蜜蜂缺乏建立抗病群体的营养物质，这导致作物的生产力迅速降低。蜜蜂生态系统服务功能的下降不能只归咎于某种人造化学物质或某种食物类型，它是在过去数十年里出现的多种因素综合影响的结果，因而不可能迅速得到解决。虽然兴旺繁盛的蜜蜂群体已经激发政策制定者和管理专家的灵感，但蜜蜂当前的状况迫使我们寻找让整个自然回归演化路径的方式和手段。尽管有机种植是改善当前状况的重要一步，我们不得不承认，生态系统中不断积累的成千上万种人造化学物质，使得该过程变得复杂而缓慢。这是一个典型案例，说明我们不能简单地只改变或禁止某种人造化学物质，而是要改变整个系统。

动手能力
Capacity to Implement

你是否见过蜂巢？蜜蜂生活在任何有花且气候允许的地方，这给我们绝好的机会，去热带、温带以及寒冷气候下观察蜜蜂的灵活性和适应性，以及它们如何生产蜂蜜、蜂王浆。问问自己：我能做什么来帮助我所在地区的蜜蜂兴旺繁盛？现在，想想你能做什么来让蜜蜂更健康，当然也包括提高它们的生产力。关键是找到帮助蜜蜂在大自然中发挥其预期功能的方式。做一些探究，每次针对一个蜂巢，致力于实现你的想法。

故事灵感来自

This Fable Is Inspired by

巴克明斯特·富勒
Buckminster Fuller

　　巴克明斯特·富勒出生于美国马萨诸塞州米尔顿市。他在学习几何学上有困难，但他高兴地用在树林中发现的材料自制工具。他推广了由德国工程师瓦尔特·鲍尔斯费尔德设计的穹顶（一种半球形的结构），这些穹顶具有连续的张力和不连续的压力。在半个世纪里，富勒有许多设计和发明，被授予许多专利。他认为社会很快将依赖可再生能源。他致力于建筑学、工程学和设计学上的能源和材料的效率问题，希望将科学原理应用于解决人类的问题。他设计了1967年蒙特利尔世界博览会的"生物圈"项目，主要由六边形小格和六个五边形小格组成。蒙特利尔市和加拿大政府后来将这座建筑物转变为一所介绍水资源和可持续发展的博物馆。

图书在版编目（CIP）数据

冈特生态童书.第三辑修订版:全36册:汉英对照 /
(比)冈特·鲍利著;(哥伦)凯瑟琳娜·巴赫绘;
何家振等译.—上海:上海远东出版社,2022
书名原文:Gunter's Fables
ISBN 978-7-5476-1850-9

Ⅰ.①冈… Ⅱ.①冈… ②凯… ③何… Ⅲ.①生态环
境–环境保护–儿童读物—汉、英 Ⅳ.①X171.1-49

中国版本图书馆CIP数据核字(2022)第163904号
著作权合同登记号图字09-2022-0637号

策　　划　张　蓉
责任编辑　祁东城
封面设计　魏　来李　廉

冈特生态童书
蜜蜂的数学
[比]冈特·鲍利　著
[哥伦]凯瑟琳娜·巴赫　绘

唐继荣　译

记得要和身边的小朋友分享环保知识哦！
八喜冰淇淋祝你成为环保小使者！